W9-AVI-088

A Pro/Manufacturing Tutorial

Pro/ENGINEER - WILDFIRE

Paul E. Funk & Loren Begly, Jr

ISBN: 1-58503-124-0

PUBLICATIONS

Schroff Development Corporation

www.schroff.com
www.schroff-europe.com

Preface

This is a tutorial manual for the beginning user of Pro/Manufacturing. The manual assumes the user has a basic working knowledge of Pro/Engineer Wildfire and CNC milling. The manual guides the user through several examples that start with the basic part model and conclude with the generation of corresponding G-code. The manual makes no attempt to teach machining techniques but does provide some helpful hints.

The manual begins with a simple click-by-click procedure to perform a single volume milling sequence for the beginner to gain a basic familiarity with the manufacturing module. The manual includes examples of hole making, volume milling, profile milling and surface milling.

The section of the manual on conventional surface milling discusses various hints and techniques for solving problems and generating procedures to develop the milling process. The manual concludes with a discussion of various Pro/Manufacturing concepts and procedures.

Pro/Engineer®, Pro/E®, Pro/Manufacturing® and Wildfire® are registered trademarks of Parametric Technology Corporation.

Copyright © 2003 by Paul E. Funk and Loren Begly, Jr. All rights reserved. Printed in the United States of America. Except as permitted under the United States Copyright Act of 1976, no part of this publication may be reproduced or distributed in any form or by any means, or stored in a data base or retrieval system, without the prior written permission of Schroff Development Corporation.

Table of Contents

Introduction

The Manufacturing module in Pro/E is used to generate machine code to produce parts. The module can be used to generate codes for CNC controlled mills, lathes, and wire EDM. All the processes consist of basically the same steps. This text will focus on generating CNC code for 3 axis milling operations for several reasons. Three axis machines are the most common and post processors for four and five axis machines tend to be very expensive. Additionally, five axis machines generally are not as accurate as 3 axis machines because of the necessary extra degrees of freedom.

There are a variety of procedures that can be used to produced the desired CNC code for a given part, but the one listed below is a logical progression for the novice user.

1) Produce the part model. The part can be produced within the manufacturing module, but generally it tends to be easier and more reasonable to produce the part model outside the manufacturing module. Beginners will especially find it much less confusing to separate the process of producing the part model and the manufacturing code. That's the method that will be used throughout this manual.

2) Create the workpiece. The workpiece represents the raw stock of material from which the part will be machined. Pro/E refers to this procedure as an assembly operation, but the workpiece can be created in manufacturing mode. The workpiece also can be created in part mode and then "assembled" with the part in manufacturing mode. In either case, Pro/E will display the original part contained within the workpiece as an assembly. In most cases, it's generally easier to create the workpiece within the manufacturing module using the part as a guide or outline. Once created and saved, the workpiece is a Pro/E part and can be modified in part mode.

3) Perform the manufacturing setup. The setup consists of defining the type of machine to use (3 axis mill, 4 axis mill, lathe, wire EDM, etc.) and the profile to be machined, the trajectory to follow, the volume to be removed from the workpiece, etc. Defining the mill volume and/or profile, etc. can also be done in the machining menu. However, it then becomes part of a given sequence and would have to be redefined if the same volume, profile, etc. was required again or if the sequence was deleted. It's analogous to defining a datum plane in Pro/E as part of a feature. The datum plane becomes part of the feature. In the same way the milling volume, etc. becomes part of a particular machining sequence and deleting the sequence would require you redefine the volume, trajectory, etc. If you define the mill volume, etc. at this step it is independent from any particular sequence and can be modified without reentering the machining menu.

The manufacturing setup also requires defining a coordinate system if one doesn't already exist and a retraction plane for the cutting tool. The coordinate system <u>must</u> match the mill orientation and the <u>part</u> zero.

4) Define the machining operations. Each operation can consist of several different sequences. Generally speaking, an operation consists of all the sequences that are performed on the workpiece without changing the setup; i.e., without reorienting the workpiece in the fixture or on the machine. Pro/E permits changing tools or machining parameters (feed rate, spindle speed, step size, etc.) for each sequence. An operation can consist of several different machining sequences (profile, volume mill, trajectory, hole making, etc.) that you perform using the same or different milling tools. If the milling volumes, surfaces, trajectory, etc. weren't created in the previous step, they can also be created at this point with the qualifications we discussed in the manufacturing setup.

5) Create the CNC code. Pro/E uses a post processor to convert the CL data (cutter location)

to codes for various mill controllers. Pro/E supplies several generic post processors to produce the

G-code. This manual will use one of those post processors. The CL code can be used by a variety

of post processors sold by vendors other than Pro/E.

NOTES:

Manual Defaults

For the first three sections of this text we will start with a basic feature that consists of a simple rectangular block. We'll need to refer to different faces of the block throughout the text. We have included here a picture of the block (in default orientation — the Pro/E menu picks, View/Orientation/Standard Orientation) with the faces named as we'll refer to them throughout the first three sections of this manual. <u>Refer back to this page whenever you need a refresher on our nomenclature</u>.

Also, **it's important the basic feature you create for the parts in the first three sections has this same default orientation**. *If it doesn't edit the basic feature (in default view) so it appears as shown here*. If you simply rotate the part to the orientation shown, you'll have trouble following some of the steps in these sections.

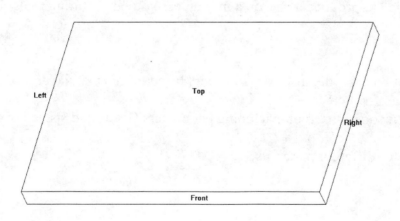

Figure A
Basic Part Orientation

This default orientation places the part in an orientation that agrees with the way the part will set on the mill. The bottom will set on the table and the top will be machined. The front of the part will face the machine operator. Pro/E's default coordinate system will not agree with a standard mill coordinate system with this orientation, but we'll define one later that will. The system we use will keep the part in the same orientation the mill uses.

UNITS

This manual will use inches as the default system of measurement unless otherwise stated. (Section 4 illustrates how units can be mixed.) We realize many companies specify metric units as the default, but most tool shops still use the English system. If someone using the manual prefers to use the metric system, all the numbers we've given will work without conversion for the *computer simulations*. Pro/E will be displaying the machining of a 4 mm by 6 mm block with a 0.25 mm cutting tool. It works in the Pro/E simulation but the CNC code generated cannot be used to produce a part on a mill unless you have cutting tools smaller than a sewing needle!!

Unless stated otherwise, the milling parameters listed in this manual are for producing a model or pattern from wood. Different machining parameters (feeds and speeds) would obviously be required for different materials.

Section 1 -- A Quick Run Through

In this section we'll develop the CNC code for a 3 axis CNC mill to produce a simple

rectangular block with a raised letter.

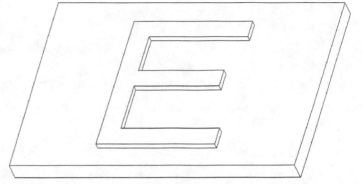

Figure 1-1
Part for Section 1

Pro/Manufacturing can quickly become very involved and complex. This exercise is designed to

work through a single manufacturing sequence to familiarize the user with the general procedure

involved. Although a single manufacturing sequence (volume milling), it still involves a lot of

commands. Since we're just trying to gain familiarity with the general procedure, we'll present this

example with a minimum of explanation. At the beginning of each step we'll explain what we're

going to do, then we'll do it and at the end of the step we'll tell you what we (and you) did. Later

sections will go into detail about why you made the choices you were told to make and what

alternatives you have available to you.

Produce the Part Model

We'll follow the steps discussed in the preface. First, 1) Produce the part model. Before

entering the Manufacturing menu, enter the Part menu and create the block shown above. A click-

by-click procedure is included in the Appendix A and B. You can name the part whatever you

wish, but we'll refer to it as "block" throughout the remainder of this section. Save the part and

Window

Close

File

New

⊙**Manufacturing**

Enter the name you want to use for your manufacturing process. We'll use "block". (Pro/E will add

a .mfg extension.) Select

OK

That completes step 1).

Assemble the Part and Workpiece

We're now ready for step 2). We'll create the workpiece material the part will be machined

from using our part as a guide. We'll make the workpiece bigger than our original part.

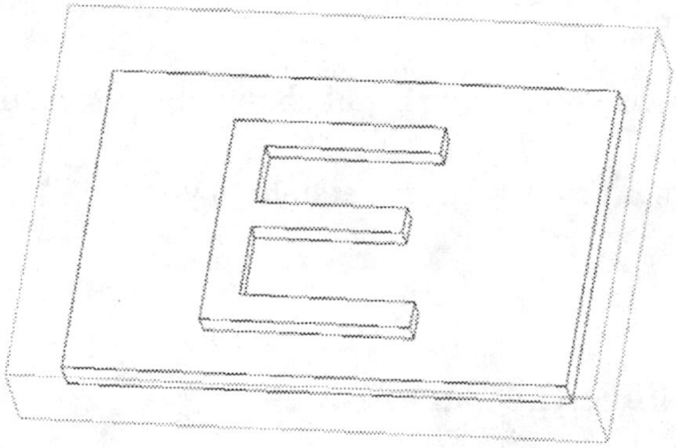

Figure 1-2
Part/Workpiece Assembly

We begin by selecting

Mfg Model

Assemble

Ref Model

and pick on the part name for the part we want to manufacture (block.prt).

Open

Pro/E displays the part. We'll use this part as a guide to create the workpiece from which the part will be machined.

Create

Workpiece

We gave it the name, block_wp. We're now going to enter Sketcher, just as if we were building a part in part mode, to model the workpiece. Select

Protrusion

Done

and select the sketching plane. Use the right mouse button to highlight the bottom surface of the block and the left mouse button to select it. (Refer to Figure A if you don't recall which is the bottom surface). Pick on the front surface (Refer to Figure A) of the block for the Reference and select

Bottom

for the Orientation. Select

Sketch

Use the left edge and front edge as references. The actual references are not important the way we'll dimension the section. Select

Close

and create a rectangular section as the outline for the workpiece so that it completely encompasses the outline of our block. Dimension the rectangle you just created to be ½ inch bigger than block.prt on each side.

Figure 1-3
Dimensioning Scheme

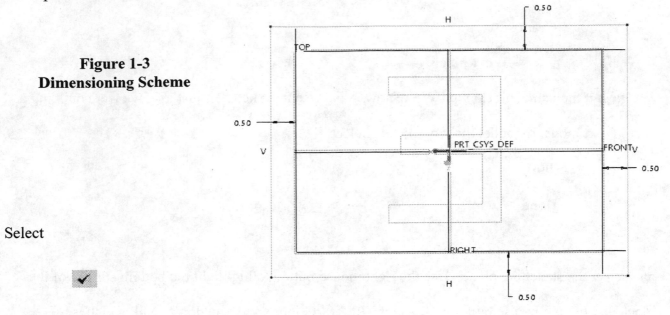

Select

Go to the default view and make sure that the protrusion encompasses the part. If it doesn't drag the depth handle or use

to change the depth direction. Enter 1.0 inch as the depth of the protrusion. Select

Done/Return

and we've completed the creation of our workpiece. In default view the part/workpiece assembly should appear as shown in Figure 1-2.

We've completed step 2). We have our original part model assembled within the workpiece from which it is to be machined.

Manufacturing Setup

We're now ready to 3) perform the manufacturing setup. We will specify a 3 axis milling operation, create a coordinate system for our operation and create the volume of material to remove by milling. Select

Mfg Setup

Pro/E displays the Operation Setup dialogue box. Click on the "mill icon" at the far right of the NC Machine box.

Make sure the Machine Type is "Mill" and the Number of Axes is "3 Axis". Select

OK

and click on the arrow next to Machine Zero. Select

Create

and pick on the workpiece. Pro/Manufacturing displays a coordinate system dialogue box. We'll place a coordinate system at the intersection of the top-left edge of the workpiece with the top front edge. Click on the **top-front** edge to indicate the location of the first reference. Hold down the control key and click on the **top-left** edge to indicate the second reference for location of the coordinate system.

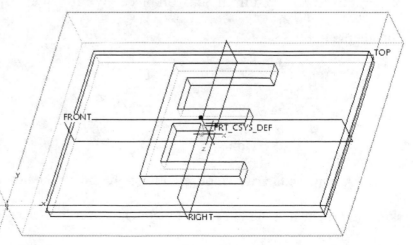

Figure 1-4
Pro/ E Axis Display

Click on the *Orientation* Tab. If necessary, flip the x axis so that it extends to the right and flip the

y axis so that it extends from the front to the back of the workpiece (Figure A).

Select

OK

OK

Figure 1-5
Part Home

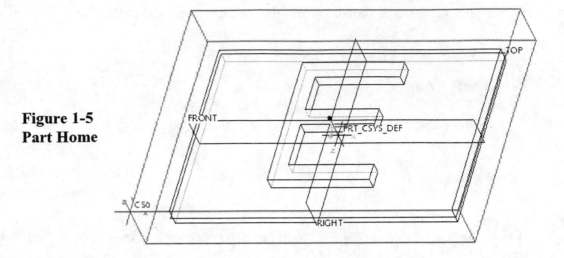

To complete the manufacturing setup, we still need to specify the volume of material to be

removed. We'll select the entire workpiece and the "remove" the part from this volume. Select

Mfg Geom

Mill Volume

Create

and enter a name. We'll use "mv1". We'll use the sketch command to create the volume of

material to be removed from our workpiece. We'll remove all the workpiece material that lies

outside the part. Select

Sketch

Done

Done

Again use the right mouse button to highlight the bottom surface of the workpiece as the

sketching plane and the left mouse button to select it. Select **Bottom** from the SKET VIEW menu and pick on the front surface of the workpiece. Again use the left and bottom edges as references.

Select

Close

Sketch

Edge

Use

and select all four outer edges of the workpiece (not the part). Select

Close

Done

Up to Surface

Done

and pick the top (Figure A) surface of the workpiece. (You may want to use the default view.)

Select

Ok

We've selected our entire workpiece as the mill volume. But we need to leave the material

that represents our part. At this point Pro/E provides a Trim function that will "trim" the part from

the mill volume. Select

Trim

and use the right mouse button to select the part we wish to be "trimmed" out of the mill volume.

Use the left mouse button to accept the selection.

Done/Return

Done/Return

Done/Return

and we've defined the volume to be removed (the workpiece minus our part).

And we've completed the manufacturing setup. We defined a 3-axis milling operation,

created a coordinate system for our workpiece, and created a mill volume representing the material

that is to be removed by machining.

Machining Sequence

Let's 4) define the machining operations. We'll select a volume milling sequence, define our

tool and machining parameters (tool size, cutting speed, etc.), create a retraction plane and specify

the volume of material to be removed (created in the previous step). Select

Machining

    NC Sequence

    Done

        Done   (notice the checked

        parameters we must define)

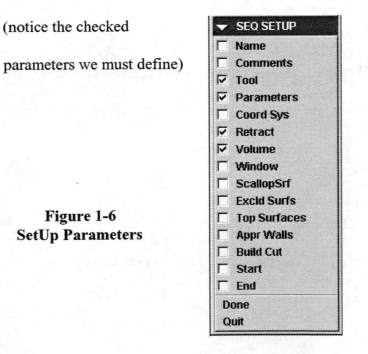

Figure 1-6
SetUp Parameters

Pro/E will display a Tool Setup Table.

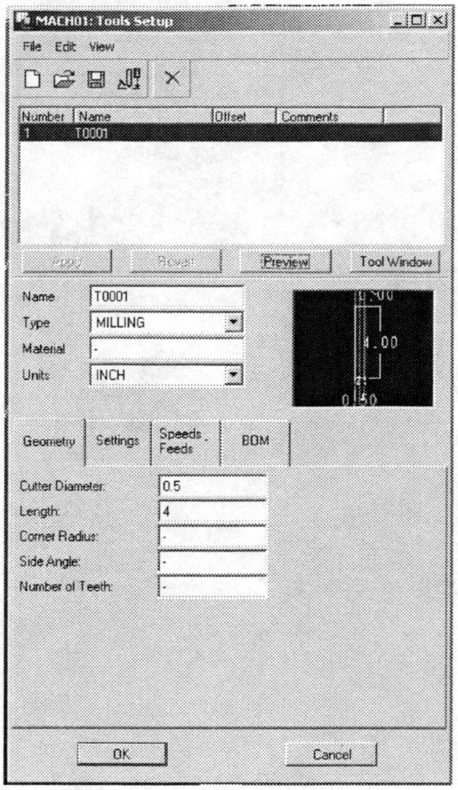

Figure 1-7
Tool Setup Table

While the user can modify any of the values, we'll just specify the diameter and length for this

example. Click on the specified parameter and change to the following values:

Cutter_Diam   .25

Length   2

To save the values and exit the Setup table,

    Apply

        File

            Done

                Set

and Pro/E will display the remaining machining parameters. While the user can modify any of the values, all parameters that have a value of -1 must be specified.

	Volume Milling
Manufacturing Parameters	
CUT_FEED	-1
STEP_DEPTH	-1
STEP_OVER	-1
PROF_STOCK_ALLOW	0
ROUGH_STOCK_ALLOW	0
BOTTOM_STOCK_ALLOW	-
CUT_ANGLE	0
SCAN_TYPE	TYPE_3
ROUGH_OPTION	ROUGH_ONLY
SPINDLE_SPEED	-1
COOLANT_OPTION	OFF
CLEAR_DIST	-1

**Figure 1-8
Machining Parameters**

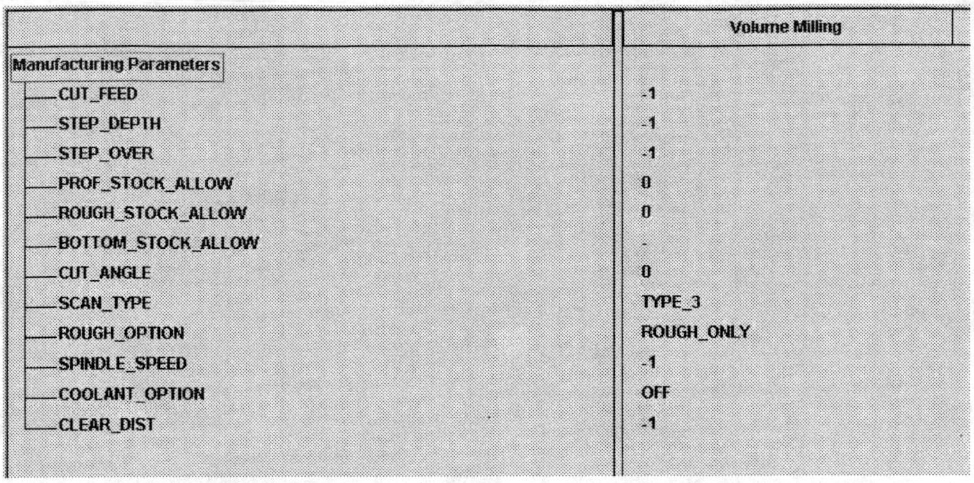

Use the following values:

Cut_Feed   60

Step_Depth   .125

Step_Over   .125

Scan_type   Type_Spiral   (We'll explain why you change this later.)

Spindle_Speed   1000

Clear_Dist   1

   After entering all the above values, select

   File

      Exit

to save the values and exit the table. Select

   Done

Pro/E now prompts us to create a retraction plane; ie, a plane to determine the height the tool will

withdraw to each time it retracts from the workpiece. We'll put a retraction plane one inch above the part (0.5 inches above our coordinate system). Select

Along Z-Axis

and enter a value of 0.5 (½ " above the workpiece).

Ok

Pro/E displays the retraction plane above the workpiece.

Figure 1-9
Retraction Plane

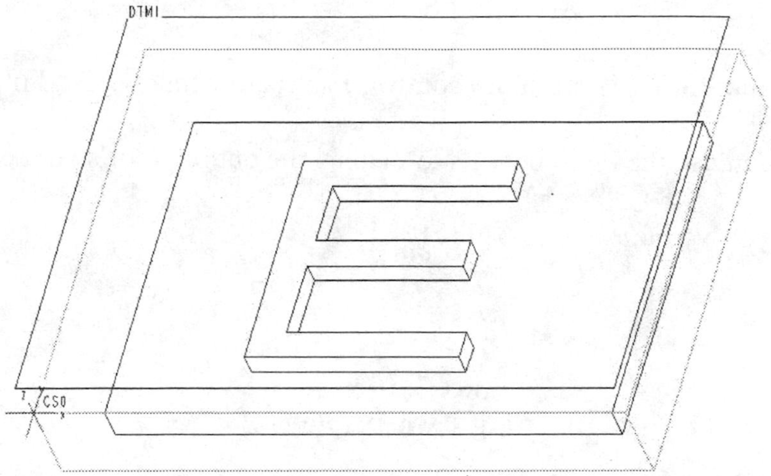

Now we must specify the volume of material to be machined. Since we've already created the volume (recall we called it mv1), all we have to do is select it. Select

Select Vol

      OK (notice the volume we created earlier (MV1) is highlighted)

and we've finished the process.

And we've finished defining our machining sequence. We defined a volume milling sequence, entered tooling and machining parameters, created a retraction plane and selected the volume of material to be removed by milling.

Viewing and Outputting Results

Although we're done, at this point we need some "proof" that everything we've done is okay. We can get that proof by creating the tool path (CL - cutter location) and viewing the cutter location as it removes the material. We'll "play the path" of the tool. Select

Play Path

> **Screen Play**

and Pro/E plays a radio control, Play Path window. And if you've done everything correctly, you can use the radio buttons to display the cutting tool's path centerline as it removes the material you've indicated to be mv1.

Figure 1-10
Cutting Path Display

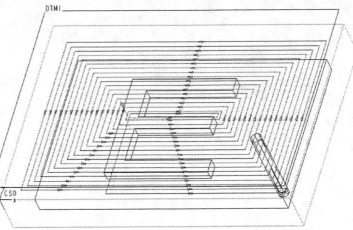

After you've played the tool path (CL data), you can also run an NC check to graphically depict the material removal. Choose

NC Check

and use Vericut controls (the green button on the bottom right of the screen) to watch as Pro/E simulates how the material is removed.

Figure 1-11
NC Check

That's everything involved in the process except posting or postprocessing the CL data to generate the G-code. That's pretty straight forward and basically just requires a listing of menu picks. We'll save that as part of a later exercise.

NOTES:

Section 2 — Holemaking, Volume Milling and Profile Milling

In this section we'll work through an example that requires hole making, volume milling and milling a profile. We'll use a part very similar to the one we used in the previous section, but we'll discuss the required selections so you'll know why we're making them. Also, in the previous section we performed an operation that contained only one sequence. In this section the operation will require three sequences; hole making, volume milling and profiling.

Figure 2-1
Part to be Machined

Assembling the Part and Workpiece

We will discuss the steps necessary to develop the G code for a 3 axis milling operation to create the plate shown. The operation will consist of a hole drilling sequence to create the holes, a volume milling sequence to cut out the three letters and a profile milling sequence to machine the plate from the workpiece. The plate is 4" by 6" with a depth of .375". The lettering was produced using a 1/8 inch deep cut. Details of the process are given in the Appendix C. (The lettering on the plate can be any lettering of your choosing. However, be aware the cutout for the lettering must be large enough to allow machining with a 1/8 inch flat end mill.) In default mode, the part should

have the same orientation as the one shown here. While not absolutely necessary, this default

orientation will make it easier to follow the procedure and discussion in this manual as well as the

Pro/E graphical representations of the milling operation. As before we'll follow the steps discussed

in the introduction to generate the CNC code to machine the plate.

First, 1) Produce the part model. Before entering the Manufacturing menu, enter the Part

menu and create the plate shown in Figure 2-1.

Give the part a name. We'll use the name "plate" to refer to the part throughout the

remainder of this section. As you work through this example, we'll need to refer to different

surfaces of the plate. You may want to refer back to Figure A to refresh your memory concerning

our face naming convention.

Retrieving the Part Model

After you've created and saved the part, you're ready to enter manufacturing mode to begin

the process which will eventually lead to the generation of the necessary CNC code. Select

File

New

⊙**Manufacturing**

Enter the name you want to use for your manufacturing process. We'll use "plate ." Pro/E

will add the .mfg extension, so our project is plate.mfg.

Creating the Workpiece

We're now ready for the second step discussed in the introduction, 2) Create the workpiece. Recall the workpiece is the raw stock from which the part will be machined. Generally, the workpiece will be larger than the part and we can use the part as an outline to create the workpiece. However, Pro/E will allow us to machine outside the workpiece. For instance, if we're milling a wood pattern, we might want to start at the outer periphery and make a light cut along the outer edge of our workpiece and machine inward to prevent the tool from "splintering" the wood outward if we were to machine from the center outward. Also, the actual workpiece that you place on the mill can be larger than what you illustrate in Pro/E. For this example the actual workpiece could be larger than what's illustrated to provide room for holding the workpiece in place.

In the manufacturing mode, Pro/E treats the workpiece/part combination as an assembly. The workpiece can be created in part mode and then assembled with the part in manufacturing mode. It is usually simpler to use the part in manufacturing mode as an "outline" to create the workpiece. That's the method we'll use in this manual. We'll begin by importing the part to use as a reference for constructing the workpiece. Begin by selecting

Mfg Model

    Assemble

       Ref Model

and select the part you previously created (plate) in part mode. Select

Open

Pro/E will display the part. We can now use it as a guide to construct our workpiece. Select

Create

   Workpiece

and enter a name for the workpiece. Pro/E will add the .prt extension. Once created the workpiece

can be modified in either part or manufacturing modes. It's also a good idea to dimension the

workpiece relative to the part rather than "overall" dimensions. In other words, in this example

rather than dimensioning the workpiece with overall dimensions of 5 inches by 7 inches, it will be

dimensioned as ½ inch bigger than the workpiece on all 4 sides. That way if the part dimensions

are modified, the workpiece will change to accommodate the new dimensions and still be ½ inch

bigger on all sides.

   This text will use the name plate_wp for this workpiece. That convention lets us know it's a

workpiece (wp) for a part named plate. We're now going to enter Sketcher, just as if we were

building any part, to model the workpiece as a protrusion that completely encloses the part. Select

   Protrusion   (This selection is equivalent to ☐)

      Done   (We'll Extrude a Solid for the workpiece)

and we're ready to select the sketching plane. For this example, we used a workpiece that was the

same thickness as the part and ½ inch bigger on all sides. The workpiece you create should

completely encompass the part. We'll select the *bottom* (Figure A) *surface of the plate* as the

sketching plane. Use the right mouse button to select the sketching plane and the left button to

accept it. The workpiece we create then will protrude from the bottom to a point on the top surface

of the part. Obviously, the workpiece could protrude from the top surface to the bottom surface.

The workpiece also could extend below the bottom surface of the part. That, however would require an additional Pro/E "operation" to remove the material extending below the bottom surface. In other words, the machine operator would have to reorient the part in the mill to remove this material and an additional Pro/E operation would have to be defined. For the sketch orientation reference, select the *front surface* (Figure A) of the part and **Bottom** for the orientation.

Sketch

Choose any two perpendicular edges of the part as references. Since we're going to dimension our workpiece with respect to the sides of the part, these references aren't critical. We'll just create a rectangular block as our workpiece, so select

□

and create a rectangular section as the outline for the workpiece so that it completely encompasses the outline of the plate. As discussed earlier, dimension the rectangle you just created to be ½ inch bigger than plate.prt on each side so that the workpiece will always be referenced to our part. Select

✔

⊥⊥   (We won't allow for any workpiece stock above the part.)

and pick on the top surface of the part. To finish creating the workpiece, select

✔

Done/Return

and return to the default view to observe how the part is "assembled" within the workpiece.

Figure 2-2
Part/Workpiece Assembly

Manufacturing Setup

We're now ready to 3) perform the manufacturing setup. We'll specify a 3 axis milling

operation and define the material to be removed. We can also create or retrieve tooling and fixtures

in the Manufacturing Setup. In this example we're not using a special fixture (to hold the

workpiece in place), and we'll define the tooling in the machining section. If the tooling was

defined here, it would simply be selected in the machining section. Likewise, we'll select the

material to be removed in this section. In the machining section then we'll just select the

appropriate material name. If we didn't define the material to be removed in this section, we could

define it in the machining section. As explained in an earlier section, defining the material to be

removed here makes it "independent" of the machining sequences.

We begin by selecting

Mfg Setup

and Pro/E displays an Operations Setup dialogue box. We can name the operation, select the type

of process (3-axis mill, 4-axis mill, lathe, etc.) We can also define the retraction plane, but we'll do

that in the machining sequence in this example.

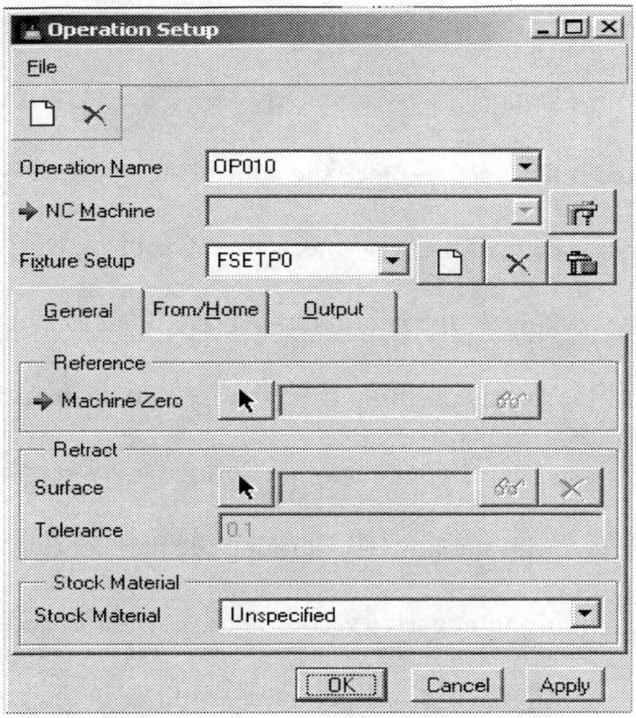

**Figure 2-3
Operations Menu**

Notice the red arrows in the Operations window indicate that we must define the type of NC

Machine we'll use in this process and the Machine Zero. We could also name the workcell, but by

default Pro/E will use the name OP010. Again, we'll wait and create the retraction plane in the

machining sequence. To continue, click on the machining icon

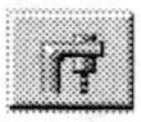

and make sure the Machine Type is Mill and the Number of Axes is 3 Axis. Select

OK

and notice the NC machine has been defined. Select the Machine Zero arrow. If a coordinate

system already exists at the appropriate location on the part or workpiece, we could simply pick it.

Since in this example, no such system exists, we'll have to create one. Select

Create

and select the workpiece (not the part) as the model we'll use to create the coordinate system. Pro/E displays a Coordinate System dialogue box and highlights the Reference for us to select up to three references to place the coordinate system. We'll place our coordinate system at the top-left-front corner of the workpiece. We want to specify the positive z-axis point upward from the top of the workpiece, the positive x-axis from left to right across the front of the workpiece, and the positive y-axis pointing from the front to the back of the workpiece. This agrees with a standard mill axis. Realize the way we defined our workpiece, the bottom surface of the workpiece will sit on the milling table. By defining the part coordinate system in the top-front-left corner, we'll be machining in negative z-coordinates. If the coordinate system is placed at the *bottom*-front-left corner, all machine coordinates would be in the positive z-direction, but it would require that we establish part home on the bottom of the part when we sat the workpiece on the milling machine. It's easier for the machinist to locate part home on the top surface of the workpiece.

To place a coordinate system on the workpiece, we'll begin by picking two edges to represent the x and y axis. (One of several methods of placing a coordinate system when using Pro/E.) The intersection of these edges will represent the origin of our coordinate system. Your first pick will select the x-axis and your second pick will select the y-axis. Begin by *picking the top-front* edge of the workpiece. Pro/E will place three perpendicular lines there indicating the temporary location of the coordinate system. While holding down the **Ctrl** key, *pick on the top-left* edge of the workpiece. Pro/E will move the coordinate system to the intersection of these two lines and indicate an x, y and z direction. Pick on the *Orientation Tab*. You'll probably have to use the *Flip* button to get the x-axis to extend to the right. The y-axis should already extend from the front to the back of the workpiece. If the direction is wrong, use the Flip button to reverse it. The third axis is defined according to the right hand rule.

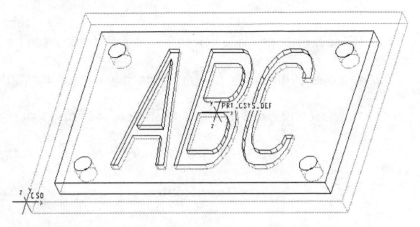

Figure 2-4
Part Coordinate System

Select

    OK

       OK

and we've completed specifying the Operation Setup. (We have yet to define the cutter, cutting

speed, etc., but as discussed earlier, we'll do that in the Machining menu as part of the

manufacturing setup.) We still need to specify the material to be removed.

Drill Group

We're going to drill four holes, machine out the letters and machine the plate from the

workpiece. We'll select the holes by defining a drill group and use a mill volume to machine the

letters. We'll define the drill group first. Select

    Mfg Geom

       Drill Group

          Create

Pro/E displays a Drill Group dialogue. Select

⊙**Pattern**   (You could leave the Single button checked and select holes individually.)

    Add

Pro/E now prompts us to select the axes of the holes to be included in the drill group. Select the axis of one of the holes. (You may need to use the axis "filter" to make the pick.) Then

    OK   (Pro/E displays a listing of the holes that we've selected.)

        OK   (Notice we kept the default DRILL_GROUP_1 name.)

    Done/Return

and we've specified the drill group. Again, this allows us to just select the group in the machining sequence. We could have waited and selected the holes in the machining sequence instead of defining a drill group here.

Mill Volume

We're now ready to define the mill volume used to machine the letters. Again, we can perform this procedure in the machining function, but we'll do it here and simply select it there. Pro/E allows the user to define a volume of material to be removed from the workpiece and then "trim" out the part. That's worth repeating. In Pro/E we can define a volume of material that we want removed (machined) from the workpiece. Pro/E then allows us several options for modifying that volume. We can add volume, remove volume, change volume dimensions, or trim. The trim function will remove from the mill volume any material that belongs in the part. That permits us to create a volume that contains a portion or all of the part and then "trim" out the part. In this example we want to mill out the letters so we'll create a rectangular volume that includes the material to be removed from the letters as well as some of the material in our plate. We'll then use the trim function to remove any of the "part material" from this volume. What remains is the material to be removed by machining. We begin by indicating we want to define a mill volume. Select

Mill Volume

    Create

and enter the name "mv1". We'll use this name as our own convention so we could have an mv1, mv2, mv3, etc. for more complicated parts and workpieces.

We'll use the sketch command and go into Sketcher to create the volume of material to be removed from our workpiece. We'll use a rectangular section which includes all of the letters. If we were to include one or more of the holes in our volume, we'd have to go back into Sketcher and add that material back to the volume. Otherwise we'd drill and mill the holes. Likewise, if our mill volume is larger than the plate, Pro/E will generate code to remove that material as well. We don't want to do that here because we're going to use a profile milling operation to mill the outer surfaces of the plate. We'll now go into Sketcher and create our rectangular volume which includes all the letters but doesn't extend to the four holes. Select

Sketch

    Done

        Done

and select the *bottom* surface of the workpiece (Figure A) as the sketching plane.

Okay

You could work from the top surface to the bottom surface of the plate instead of the method we're using. Remember we're going to trim out any volume that includes our original part (plate.prt) anyway. Make sure Bottom is highlighted as the orientation reference and click on the front surface (Figure A) of the workpiece.

Select two perpendicular edges of the part as references. The actual references chosen are not critical. Select

Close

▢

and create a rectangular section that includes all the letters but <u>does not include any of the four holes</u>. The actual dimensions of the section are not critical since we'll "trim" out all the material that belongs in the original part leaving only the material in the letters to be machined away. Select

✔

Up to Surface

Done

and pick the top surface of the workpiece. (You'll have to use the standard orientation to see the top of the workpiece.) Select

OK

We've created a mill volume that includes some of our part. (It will be highlighted on the screen.) We need to leave any material that represents our part. This is the point where we use the trim function to remove the part from the mill volume. Select

Trim

and pick our reference part. In the message window, Pro/E indicates the trim function has been performed, but let's check to be sure. Select

Done/Return

   Shade

and make sure MV1 is highlighted. Select

OK

Pro/E will display the volume (MV1) to be milled after the trim function was completed.

Figure 2-5
Mill Volume

Select

   Done/Return

   Done/Return

and we've defined the volume to be removed (the mill volume we created in Sketcher minus our part). For more complicated parts we could now create additional volumes or add to or subtract from mv1.

That completes the manufacturing setup. It's much easier and less complicated to define the profile surfaces in the machining operation.

Defining the Machining Sequence

Now let's perform step 4), define the machining sequences. Recall again that in the introduction we stated that there could be several sequences within an operation. Everything we're doing in this example is part of the same operation. In the machining function we'll create three different sequences; hole making, volume milling and profile milling. There's no need to reorient the part on the mill and create an additional operation. The hole making sequence will drill the four holes, the volume milling sequence will remove the material representing the letters and the profile milling sequence will machine the four perimeter surfaces of the plate.

The Hole Making Sequence

Select the machining function by picking

Machining

If you select, Operation, you'll notice the current operation we're working on is OP010. We don't want to change that. To create our first sequence in the current operation, the hole making sequence, select

NC Sequence

    Holemaking   (Be sure to change this from Volume.)

        Done

and notice we can countersink, ream, etc. We'll leave the Drill and Standard options checked and select

Done

Figure 2-6
Hole Making Menu

Notice the checked boxes indicating parameters we must define. For the "Holes" menu item we'll be able to select the drill group we defined previously. Select

Done

If we had previously defined our cutting tool (drill) in the manufacturing setup, we could simply retrieve it here. Since we didn't, we must "set up" the tool parameters. Pro/E displays a Tools Setup table with some predefined values. We'll modify only those values that suit our needs. Click on the specified parameters and change their values to:

Cutter_Diam   .375   (Since we're drilling a 0.375 inch diameter hole.)

Length   2

Selecting

Apply

File

Done

will exit the table and save the values. Notice you can "Save" the tool for retrieval in another sequence. We don't need to do that here.

Now we need to define the remainder of the manufacturing parameters: feed rate, cutting speed, etc. Select

Set

and ProE will display the remaining user defined parameters. While the user can modify any of the values, all parameters that have a value of -1 must be specified. (Pro/E will not let you continue to the next step if you don't specify all these values.) Leave the Scan_Type as SHORTEST. We'll discuss this concept and all the parameters more fully when we set up the volume milling sequence.

Use the following values:

Cut_Feed   40

Spindle_Speed   1500

Clear_Dist   1

After entering all the above values, select

File

    Exit

to save the values and exit the table. Select

Done

Pro/E now prompts us to create a retraction plane; ie, a plane to determine the height the

tool will withdraw to each time it retracts from the workpiece. We'll put a retraction plane one half

inch above the part. Select

Along Z Axis

and enter a value of 0.5 (½" above the workpiece). Select

OK

Pro/E displays the retraction plane above the workpiece and a Hole Selection dialogue box. Select

Depth

    ⊙**Thru All**

    OK

Since we have already defined the holes in a "drill group", we can just select that group. If we had not previously defined the drill group, we would select the holes here. Pick on the "Groups" folder and select

Add

DRILL_GROUP_1

OK

OK

Done/Return

Viewing the Results

Although we're done with the drill sequence, at this point we need some "proof" that everything we've done is okay. We can get that proof by creating the tool path (CL - cutter location) and viewing the cutter location as Pro/E displays how the material will be removed. We'll "play the path" of the tool. Select

Play Path

Screen Play

and if you've done everything correctly, Pro/E will display a PLAY PATH window with radio buttons that are used to display the cutting tool's path as it removes the material that you've indicated to be the drill group, DRILL_GROUP_1.

After you've played the cutter path, you can also run an NC check to graphically depict the material removal. Choose

NC Check

and you can watch as the Pro/E display opens Vericut to allow you to view how the material is removed. And since we've finished with the drill sequence and are ready to create the next

machining sequence, select

Done Seq   (Saves the sequence we just created. Quit Seq will abort the sequence.)

and we're ready to begin defining the milling volume sequence.

Volume Milling Sequence

Recall the volume milling sequence will remove the material representing the three letters.

We have to define a new NC sequence. From the Machining menu, select

NC Sequence

If we wanted to go back into the holemaking sequence to view it or make changes we'd select that

sequence. However, we wish to define a volume milling sequence, so select

New Sequence

and, since Volume milling is already highlighted, select

Done

Note that Tool, Parameters, and Volume are checked. If we wanted to define a new retraction

plane, we could check that box as well. We'll define a new cutting tool, new parameters (cut feed,

speed, etc.) and specify the volume to be milled. Recall we've already created it and named it mv1.

We'll use the same retraction plane, so leave that unchecked. Select

Done

and Pro/E again displays the Tool Setup table.   Notice Pro/E has changed the tool Name for us.

That won't always happen as we'll see in the next section.

For this example, we used:

Name   T0003

Cutter_Diam   .125

Length   1.0

The two parameters we have defined should be pretty obvious. The diameter and length refer to the diameter and length of the cutter respectively. The cutter diameter must be small enough to fit within the spaces in the letters or Pro/E will not remove that section. Likewise the length must be great enough to reach the dept of any pockets. Notice we can also change the units. We can also place a radius on the tool if one was required. (We'll do that in Section 4 to create a tool for milling a contoured surface.)

Select

      Apply   (Notice that Pro/E assigns a new pocket "Number" for the tool)

      File

      Done

and the new tool has been created (T0003). We again need to set the parameters for this tool (cutting speed, etc.) Select

    Set

Again, Pro/E displays a table of parameters in which all the -1's must be changed. You'll notice in the upper right hand corner of the table there's a button labeled "Advanced". We'll do some things to illustrate the use of these parameters in Section 4 of this text. For now just click on the Advanced button to observe the available options. Click on Simplified to return to the original menu.

We'll again change all the values with a -1. The cut feed is the feed rate in inches/minute

(unless we change that parameter in Advanced). The step depth is the z-increment for the cutter

when the tool's z-location changes. Pro/E will compensate if the increments do not work out to an

even number of steps to reach the required depth. The step over is the lateral amount the cutter

"steps over" after each pass. The step over must be less than or equal to the cutter diameter. A

good rule of thumb is to make the step over no greater than 75% of the cutter diameter. We'll

illustrate the use of the "Cut Angle" in Section 4 of this manual.

Notice also there's a "Scan_Type". This parameter indicates how the cutting tool

progresses around or across the workpiece. In the table, highlight **Type_3** and press function key

F4. Pro/E displays a drop down menu that includes all the available types of milling paths for this

sequence. They each indicate how the tool progresses across the workpiece. For instance, for

Type_1 the cutting tool starts along one edge and makes successive adjacent passes and rises above

obstructions (section where material is not to be removed). For Type_2 the tool moves around

obstructions and Type_3 reverses the path at an obstruction.   In this example, we selected

Type_Spiral where Pro/E starts at different regions and "spirals" outward.

The spindle speed should be obvious. It's units are in rpm unless the Advanced parameter

has been modified. The "clear_dist" represents the distance above the workpiece at which rapid

motion ends and plunge feed begins. The plunge feed rate is the same as the cut_feed unless

changed in the Advanced parameters.

We'll use the following parameters for this example:

Cut_Feed   100

Step_Depth   .125

Step_Over	.12
Scan_Type:	Type_Spiral (You can use F4 to select this parameter.)
Spindle_Speed	2000
Clear_Dist	1

      File

            Exit   (Saves the parameters we just entered.)

            Done

Specify Mill Volume

Now we're ready to specify the volume to be machined. Since we've already created the volume (recall we called it mv1), all we have to do is select it. Select

      Select Vol

Make sure MV1 is highlighted and select

      OK

and we've finished the volume milling sequence.

Viewing the Results

To again view the results, select

      Play Path

          Screen Play

and you can use the radio buttons to review the results. Again you can perform an NC check with

      NC Check

and if you've specified a cutting tool smaller than the sections of material to be removed, Pro/E will display the results. Notice material remains in some of the "square" corners where our tool could not reach.   We could use another sequence with a smaller tool to finish this area but there's always going to be some radius here. We've finished the volume milling sequence, so

Done Seq   (Exits and saves the current sequence.)

Profile Milling Sequence

As soon as we finish the profile sequence, we'll be able to generate the NC code for the entire operation. To complete the last sequence (profile milling), select

NC Sequence

New Sequence

Profile

Done

and since we're going to use a different cutting tool, check the *Tool checkbox* in addition to the items (Parameters and Surfaces) that are already checked.   Select

Done

Notice the cutting tool parameters appear the same as with volume milling. We'll also have to change the tool Name or we'll simply modify the existing tool we used for the volume milling sequence. You can change the Name to T4 or T0004, etc. The only limitation is that the Name must start with a T (It doesn't have to be a capital T.). We used the following parameters:

Name   T0004

Cutter_Diam   .375

Length   2

Click on the *Settings* tab and enter **3** for the Tool Number so that it gets placed in a different pocket.

Apply (Pro/E adds another pocket with the new Tool Name.)

 File

  Done

   Set

and Pro/E again displays the remaining machining parameters. We used

Cut_Feed   400

Step_Depth   .25

Spindle_Speed   2000

Clear_Dist   1

 File

  Exit

  Done

and Pro/E indicates for us to pick the surfaces for the profile milling operation. Notice we can select them on the model, the workpiece, from a mill volume or mill surface (if we had previously defined one). We want to pick our surfaces on the model, so leaving "Model" highlighted, select

Done

and select the surfaces you wish to mill. (You may have to Repaint the screen to view the surfaces.) After picking the first surface, you'll have to use the **Ctrl** key to pick additional surfaces. Pick the four surfaces that define the outer perimeter of the part (front, back, left, right). Be careful not to accidently select the top surface.   Use the right mouse button to highlight the surface and the left

mouse button to select it. When finished, select

    OK

        Done

            Done/Return

                Play Path

                    Screen Play

                        NC Check

and we've finished defining our three sequences.   In the Profile milling sequence, Pro/E just mills

the surfaces we select with the diameter cutter we specify. Unlike volume milling, it may remove

all the material or leave some, depending upon the cutter diameter.   Select

    Done/Return

        Done Seq

and we're ready to output our NC code.

                            Figure 2-7
                 The Finished Product

Creating the CNC Output

Now that everything appears to be the way we want it, we're ready for step (5). We'll use one of the generic postprocessors built in to Pro/E to generate our CNC code. We'll create the Machine Control Data (MCD) file. (The MCD file is an ASCII file containing the necessary G-codes.) It will have a .tap extension. Pro/E generates cutter location files (ASCII format) that must be post processed to create the MCD file. Select

CL Data

> **Output**

>> **Operation**

If you wanted to output the CNC code for a specific sequence, you could do it at this point by selecting "Sequence" instead of "Operation". Pro/E will list all the sequences you have created and you could select the one you want. Since we're interested in generating the code for all the sequences we've created (and in the order we created them), we select Operation. Select

OP010

> **File**

and put a check in the checkbox for MCD file. You can uncheck the CL File if you wish. We don't need to output the CL data file for this example. Select

Done

and enter a name for the MCD file (block). Select

Done

and Pro/E displays a listing of the available post processors. Select the one that most closely fits your controller. (As you place the cursor on each item of the PP List, Pro/E displays the controller name in the message window. We've found that with some editing several of the choices can be used for the 3-axis mill with an Anilam controller that we use.) For our purposes here, we'll select

UNCX01.P12

Pro/E will display a summary of the postprocessing (Interactive mode). When the process is finished you will have an MCD file (block.tap). You can review and close the information window.

The Anilam controller that we use requires that we do some editing of the .tap file created by the post processor. Pro/E places a message at the top of the file that the Anilam controller won't accept. Any text editor can be used to delete the message. There may also be some lines at the end of the file that need to be deleted.

CAUTION: The generated CNC code may move the cutting tool from the home position to the start position without raising the tool. To prevent tool breakage, edit the file to include a G00 move in the positive z-direction before the first G00 move to new x-y coordinates. We add a G00 Z0.5 at the appropriate location in the program. The Anilam controller also requires that the .tap file be saved with a .G extension.

Section 3 — Trajectory and Surface Milling

In this section we'll use the same basic feature we used in the previous two sections. We will sketch two curves on our basic feature and we'll use a trajectory milling sequence to cut along the curves. In the next section we'll work with more complex surfaces, but to gain some experience working with a surface milling sequence, we'll mill a flat surface in this section. Our part will again be .375 inches thick, and we'll machine it from a ½ inch thick workpiece. We'll use a surface milling sequence rather than volume milling to remove the extra thickness. We're going to be much less descriptive of many of the commands required in this section. If you need additional instruction, referring back to the previous sections should provide sufficient directions.

Create the Part

Create the part shown. It's a 4" by 6" by .375" plate with the two datum curves sketched on the top surface. A step by step procedure to create the part is given in the Appendix D.

Figure 3-1
Part for Section 3

After you've created the part and saved it quit the window and enter the manufacturing menu. Enter a manufacturing name, enter the manufacturing module and retrieve the part you created to be used as a reference model.

Part/Workpiece

Create a workpiece that is the same size as our part but **extends 1/8 inch above the top surface** of the part. Don't forget to make sure the workpiece encompass the part. (The workpiece will be 0.5 inches thick.) We'll remove this extra 1/8 inch of material using a surface milling sequence. (We'd have to hold the workpiece/part in a milling vise for this example.)

Figure 3-2
Part/Workpiece Assembly

Manufacturing Setup

Since we'll create a surface in the manufacturing setup and we haven't done that before, we'll step through the procedure. We're going to create a coordinate system just as we did in previous sections. Then we'll define the surface to be milled. The surface must be flush with the top surface of our part. Setting up the coordinate system is the same procedure developed earlier. We won't include all the steps and discussion here. To start, select

Mfg Setup

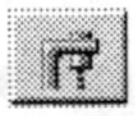

OK

Select

Machine Zero

> **Create**

and select the workpiece as the model to create the coordinate system in. Create the coordinate system at the front-top-left corner as in the two previous sections. Pick the top front edge for the x-axis and (Ctrl key) the top left edge for the y-axis. Use the orientation tab to flip the directions.

After creating the coordinate system, choose

OK

> **Mfg Geometry**

> > **Mill Surface**

> > > **Create**

and enter a name for the mill surface. (We used sf1). Choose

Add

> **Copy**

> > **Done**

and select the top surface of the part (not the workpiece) as the surface to add. In the machining menu, we'll select this surface we're creating to machine. If we create the surface on the top surface of the workpiece, there will be no material to machine away. Pro/E creates a surface on the top face of our part. Select

> **OK**   (from the SELECT menu)

> > **OK**   (from the SURFACE:Copy reference box)

> > > **Done/Return**

and we've created surface sf1 for machining.

Machining Sequences

Continue selecting Done/Return until you return to the original Manufacturing menu. Select

Machining

    NC Sequence

Conventional Surface Sequence

We'll first remove the extra material on top of our workpiece using a conventional surface

milling sequence. Select

Surface Mill

    Done

        Done

and enter the following parameters in the Tool Setup:

Cutter_Diam   0.5

Length   2.0

Apply the parameters and exit Tool Setup. Choose

    Set

and enter the machining parameters:

Cut_Feed   500

Step_Over   .40

Spindle_Speed   2000

Clear_Dist   1.0

Exit the Parameters Tree and select

Done

and create a retraction plane located with respect to the part coordinate system (Along Z Axis) 0.5

inches above the workpiece.

Figure 3-3
Retraction Plane

After creating the retraction plane, select

Mill Surface

Done

Select Srf

(If we hadn't created the mill surface as part of the manufacturing setup we could do it here.) Select

the surface we previously created by selecting

SF1   (The surface will be associated with a Quilt in the Item description.)

OK

Okay

Select All

Done/Return

OK

Use Play Path and NC Check to view the results. Notice that with surface milling the entire surface is removed. Pro/E doesn't leave small sections of material in the corners as with volume milling. With volume milling, Pro/E stays within the specified volume. For surface milling, Pro/E machines the entire specified surface even if the tool must move off the workpiece.

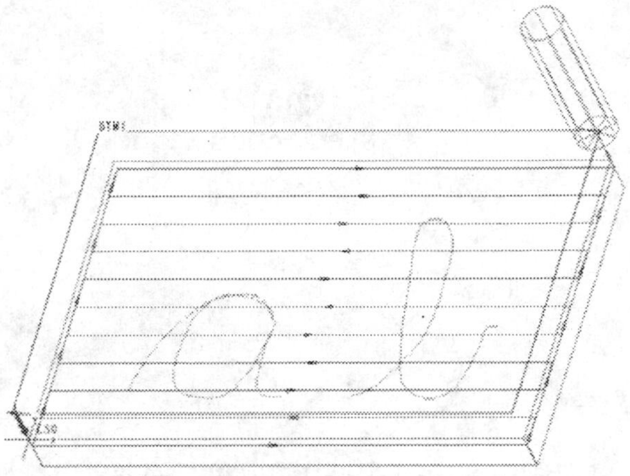

Figure 3-4
Tool Path for Milled Surface

Trajectory Milling Sequence

We'll next create the trajectory milling sequence. That follows the procedure used in previous sections, but we'll step you through most of it. Return to the Machining menu and select

    Machining

        NC Sequence

            New Sequence

                Trajectory

                    Done

and check the Tool checkbox since we'll use a different tool for this sequence. Select

Done

and enter the parameters:

Tool_ID	T0002	(Or enter a tool number higher than the listed one.)
Cutter_Diam	.125	
Length	2.0	

Click on the Settings tab and **change the Tool Number to 2**. Apply the new parameters and exit the Tool Setup. Select

Set

and enter the parameters:

Cut_Feed	100
Spindle_Speed	1500
Clear_Dist	1.0

Notice the option for Step_Depth does not contain a -1. That's because the trajectory will simply follow the curve we select. We'll sketch our datum curve on a plane 1/8 inch below the top surface of our part so the trajectory will be milled at a 1/8 inch depth into our part. Exit the manufacturing parameters menu and select

Done

Follow Sketch	(From the "drop down" menu)
Sketch	(From the Follow Sketch menu)

Use Make Datum to create a datum plane for the sketching plane. The datum plane should be offset .125 inches below the top surface of the part. To create this datum plane, select

Offset

and pick the *top surface* of the part. Select

Enter Value

and enter the offset value (-.125). The negative value creates the datum plane below the top surface

of the part.

Figure 3-5
Offset Datum Plane (DTM2)

We'll "sketch" the required trajectory on this plane. Pro/E will create a tool path which follows the

trajectory on this plane. Select

Bottom

and pick on the front surface of the part or workpiece for orientation. Select references and then

Sketch   (from the Main Menu – top of the Pro/E window)

Edge

Use

and pick on the first curve (one curve only). Select

OK   (from the Follow Sketch dialogue box)

and we've completed specifying the trajectory for the first curve. Notice that Pro/E lists "1. Follow

Sketch" as the first operation in the Customize listing.

Since we're going to use exactly the same tool and machining parameters for the other cut, we can specify it here as well. You might find it easier to just finish this NC Sequence and begin another one for the second letter, but we'll illustrate how you can include the other letter here. With the "Follow Sketch" still selected in the drop down menu, select

Insert   (opens the "Follow Sketch" dialogue box again)

    Sketch   (from the "Follow Sketch" dialogue box)

and again use

Make Datum

to create an offset datum plane to use as the sketching plane. (The "Use Prev" will use the sketching plane from the previous feature; not the current one.) Again create the plane .125 inches below the top surface of the part. Then select

Bottom

and again pick on the front face of the workpiece. Select

Sketch   (from the Main Pull Down Menu–top of the Pro/E window)

    Edge

        Use

and select the second curve. Select

    ✔

    OK   (from the Follow Sketch dialogue box)

At this point you could play the path and perform an NC Check, but you'll discover there's a problem. The tool doesn't retract from the workpiece between the end of the first letter and the beginning of the second. The tool cuts through the workpiece as it moves between the two letters.

We need to insert some more operations to correct the problem. Highlight

2. Follow Sketch

in the Customize listing and from the drop down menu select

Retract

OK (from the Retract menu)

We've now retracted the tool from the workpiece, but we need to plunge back into the workpiece to

continue machining. From the drop down menu, select

Plunge

OK   (from the Plunge menu)

While we're editing the sequence, we also need to adjust the start of the machining process. As

currently defined, the cutter will move from the end of the surface cut straight to the beginning of

the first trajectory. We don't want that to happen. The cutter should move directly above the start

point of the first trajectory, plunge cut into the metal and then machine the trajectory. So we need

to add another plunge cut. Highlight

1. Follow Sketch

and from the drop down menu select

Plunge

OK

Notice the Customize listing should now contain:

1. Plunge

2. Follow Sketch

3. Retract

4. Plunge

5. Follow Sketch

which is the sequence we need to perform to machine the letters.

We're ready to play the cutter location path and perform an NC check to confirm your results. But you'll notice with the NC check it's a little confusing. It appears that the tool is cutting 1/4" deep — through the surface that will be milled away and 1/8" into our original part. When we view the entire operation that won't be a problem.

Looking at the Results

We can perform an NC Check for the entire operation (including all sequences) by saving the CL data file and then performing the NC Check. Use

Done Seq

and return to the original Machining menu. Select

CL Data

　　Operation

　　　　OP010 (Select the current operation; not individual sequences.)

　　　　　　File

　　　　　　　　Done　　　　(leave the MCD file unchecked unless you

　　　　　　　　　　　　　　　　wish to output the G-codes at this time.)

and enter a file name for the CL Data file. Pro/E will create the file. Then select

Done Output

> **NC Check**

>> **Cl File**

and select the file name for the CL Data just created.

> **Open**

>> **Done**

Pro/E will use Vericut to display the entire operation.

Figure 3-6
NC Check of Finished Product

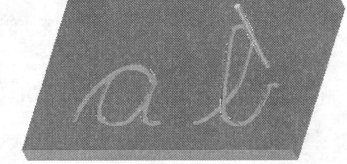

Select

> **Done/Return**

At this point (assuming we're satisfied with the results so far) we can create the G-code file.

We can output the MCD file (CNC code) as before if we select

> **CL Data**

>> **Output**

>>> **Operation**

>>>> **OP010**

>>>>> **File**

and check the MCD File checkbox and proceed as in the previous section.

Section 4 — Machining a Part with Complex Surfaces

In the first three sections of this manual, we were more concerned with teaching you the "mechanism" for producing the cutter path and CNC code for a manufacturing process. In this section we'll give you a rather simple part but will produce a more usable machining operation. The part model will be a car that we will machine out of wood. The part model will have dimensions of millimeters. Our machine parameters will be a mixture of English and metric units. In particular we'll specify several of the machining parameters in inches per minute. The final product will require a very small step_over increment for the cutting tool. We'll look at methods for rapidly viewing the results before we set the parameters for our final finished surface. We'll also look at some machining techniques to produce a better finish.

Model the Part

The part model for this section is shown below. We named it car.prt. Appendix E contains all the steps necessary to produce the part model. Be sure the default units are set to millimeters for this example in the manufacturing and part modes.

Figure 4-1
Design Part

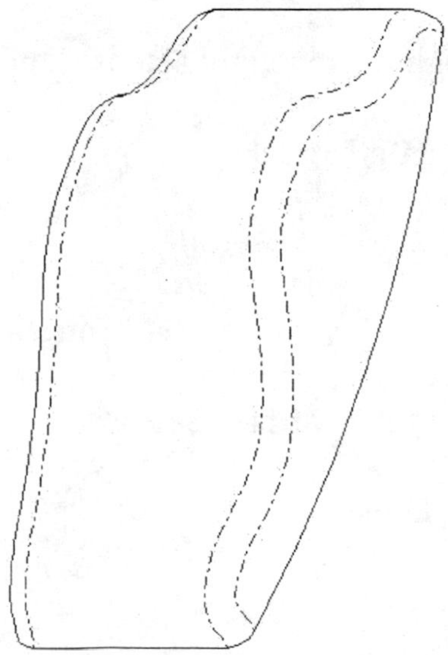

Create the Workpiece

Our workpiece will consist of a wooden block that is 360 mm long by 160 mm wide by 110 mm high. We'll use a volume milling sequence with a flat end mill to remove the bulk of the extra material and a conventional surface sequence with a ball end mill to produce the final finish. Realize that we can't just hold down our original workpiece in a conventional manner as in the previous sections. After the volume milling sequence is completed, only the rough shape of the car will remain and all the surrounding workpiece will have been machined away. The workpiece will have to be secured to a backing plate by some mechanical means such as placing screws up through the backing plate into the car body. We won't show that fixture in our manufacturing setup but it's something the machinist must deal with.   Select

File

New

⊙**Manufacturing**

and enter a name for the manufacturing process. We used car (.mfg).   Select

OK

Although the remaining steps should be familiar to you, we'll list them here. Select

Mfg Model

Assemble

Ref Model

and highlight car.prt. Select

Open

Create

Workpiece

and enter a name for the workpiece (car_wp). Select

Protrusion

Done

and select Top as the sketching plane. Flip the sketch view direction if it doesn't encompass the part. Select the Front datum plane for the Sketch Orientation and select

Bottom

for the Orientation. Select

Sketch

Pick the Right and Front datum planes as references. Create a rectangular section that completely encloses the design part. Make the workpiece about 360 mm long and 160 mm wide and centered about the car shape. You can make the workpiece bigger if necessary to completely enclose the reference model. Choose

and enter a depth of 110. (If this isn't enough depth for your sketch, you can modify the dimension or drag the depth until the workpiece completely contains the reference model.) Select

Done/Return

and we're ready for the manufacturing setup.

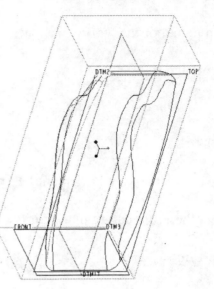

Figure 4-2
Part/Workpiece Assembly

Manufacturing Setup

In our manufacturing setup we won't define the mill volume or the surface to be milled.

Instead, we'll define a "mill window" and define our coordinate system (part home). For the

milling sequences, we'll show you how to use the "window" menu pick in the Machining section to

select the mill volume and surfaces. We'll define a "mill window" on the top surface of our

workpiece that includes the entire workpiece. Think of the workpiece as being transparent and the

part as being solid. If you look down through the mill window, you'll see all the volume of material

in the workpiece located around and above the car. Pro/E will select that material as the mill

volume. Likewise, you'll notice that all the outer surfaces of the car (excluding the bottom) are

visible from this window. Pro/E will select all those surfaces as our mill surfaces. Any "hidden"

volume or surface will not be selected. It's an easy way to select surfaces and volumes for milling.

We'll define our coordinate system before defining this window. Select

Mfg Setup

Ok

Machine Zero

Create

and use the procedure discussed in the previous sections of this manual to establish a coordinate

system in the top-left-front corner of the workpiece. (Corresponding to the actual top-**right**-front

corner of the car.) Make the x-axis point across the width of the car from left to right on the

workpiece, the y-axis point along the length of the car from the front to back and the z-axis point

upward, away from the top surface of the car (See Figure 4-3).

Figure 4-3
Part Coordinate System

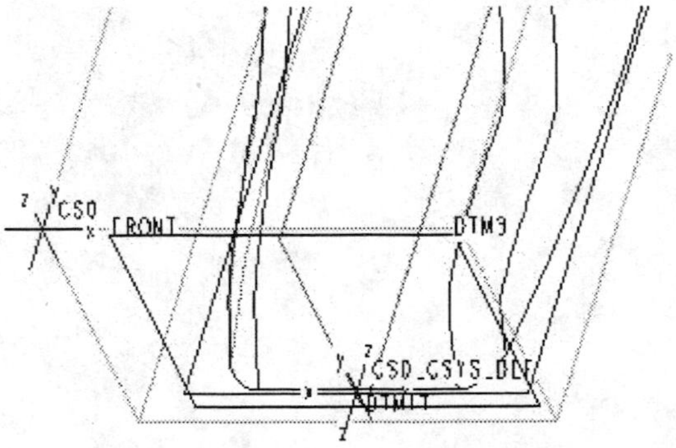

When finished, select

OK

and we're ready to create the mill window that we'll use to select our mill surface and volume.

Creating a Mill Window

Select

Mfg Geometry

Mill Window

      Create Wind

and enter a name for the window (MW1). Select the coordinate system we just created on the

workpiece, not the default coordinate system. Pick on the top surface of the workpiece to define the

window plane. Choose

   Select

and pick on each of the top edges of the workpiece (Use the Ctrl key). Select

   Done

      OK

and we've defined our mill window, and we're ready to enter the machining menu.

Figure 4-4
Part with Mill Window

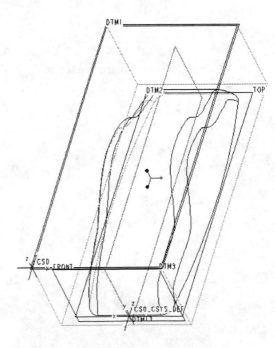

Defining the Machining Sequences

In the machining menu we'll define the cutting tool, machining parameters, volume milling sequence and a conventional surface milling sequence. But we'll need to go into the advanced machining parameters since we're working with a combination of units. And we'll use the window menu selection to select the volume and surfaces to machine. To produce the final surface finish it will be necessary to specify a very small step_over. But that requires a great deal of time to visualize. We'll use a larger step_over and a much larger "increment" so we can rapidly see what's happening. After everything is working to our satisfaction, we'll go back and modify these values to generate our final code. Select

> **Machining**
>
>> **NC Sequence**
>>
>>> **Done**   (leave Volume highlighted)

Mill Volume Sequence

<u>Check the checkbox for Window</u>. Notice when you place a check in front of Window, the Volume check disappears. That's because Pro/E will select the Volume by using the window we defined in the manufacturing setup. Pro/E will automatically pick all the volume for us. After placing the checkmark, select

> **Done**

and enter the following parameters (notice the length units are now in millimeters).

Cutter_Diam   12.7

Length   120   (May have to be longer if your car is too tall.)

Apply the parameters and exit the Tool Setup.

Advanced Machining Parameters

Since we're using a mixture of English and metric units, we'll have to use some of the

advanced machining parameters. Select

Set

Advanced

Under CUT OPTION, enter

Scan_Type   Type_Spiral   (Use F4 or drop down menu to make a pick.)

and under CUT PARAM enter

Step_Depth   20   (We'll modify this to be 10 for the final output.)

Step_Over   12   (We'll modify this to be 8 for the final output.)

Rough_Stock_Allow 1   (Leave 1 mm of stock for the finish cut.)

Prof_Stock_Allow   1

Under FEED, enter

Cut_Feed   500

Plunge_Feed   50

Ramp_Feed   10

Cut_Units   IPM   (Use F4 or drop down menu)

Retract_Units   IPM   (Use F4 or drop down menu)

Plunge_Units   IPM   (Use F4 or drop down menu)

and under MACHINE, enter

Spindle_Speed   2000

and under ENTRY/EXIT

Clear_Dist   50.8

    Select

    Simplified   (Make a quick check to be sure you have entered all the required

                         parameters)

and exit the machining parameters menu. Select

    Done

        Along Z Axis

and enter a value of 15 (Recall we're working in millimeters.).

Creating the Mill Volume

    We'll use the mill window we created to select our mill volume. Choose

    Select Wind

        Find Now

and select the window we created previously. Select

    MW1 (The window description will have a feature associated with it.)

        Ok

and Pro/E selects all the eligible volume in the window we've defined. We can play the path to

observe the results. Select

Play Path

   Screen Play

and you'll probably want to adjust the speed using the slider in the dialogue box to make the

process go quicker. You can use Rewind to reverse the process and a slower setting for the forward

direction if you'd like to more closely observe what is happening.

   Close

Perform an NC Check to better visualize the results. You'll notice we've made a pretty rough

cutout of our car. We can improve this by modifying our step over and step depth.

Modifying Machining Parameters

   We need to remember we are machining this part out of wood. If we start on the inside and

spiral outward, the wood will tend to "split out" as we approach the edge. We need to change that.

Return to the NC SEQUENCE menu and select

   Seq Setup

and place a check in front of the Parameters checkbox. Select

   Done

      Set

         Advanced

and under the heading of Cut Option you'll find a parameter called

Cut_Direction   Standard

click on Standard and use F4 to select Reverse. Exit the Parameters menu and play the path.

Notice Pro/E begins at an outer edge and spirals inward. You can slow down the speed or perform an NC Check if you have difficulty observing this.

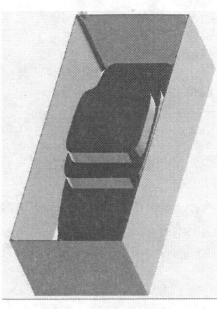

Figure 4-5
Reversed Cut Direction

Surface Milling Sequence

After viewing the results and ending the sequence (Done Seq), return to the MACHINING menu. Select

NC Sequence

> **New Sequence**

> > **Surface Mill**

> > **Done**

Now we want to use a different tool to machine the surfaces, and we'll use the window we created to specify the surfaces. Check the checkbox for both Tool and Window. Again notice when you place a check in front of Window, the Surfaces check disappears. That's because Pro/E will select the surfaces using our window. After placing the two checks, select

Done

and modify the tool parameters to the following:

Tool_ID	T0003	(Must be changed or you'll simply modify T0002.)
Cutter_Diam	12.7	
Corner_Radius	6.35	(One half the tool diameter)
Length	120	

Use the Settings tab to increase the Tool Number. Apply the parameters and exit the Tool Setup.

Advanced Machining Parameters

Select

Set

Advanced

Under CUT OPTION, enter

Scan_Type	Type_3	

and under CUT PARAM enter

Step_Over	12	(We'll modify this to be 1 for the final output.)
Cut_Angle	45	

Under FEED, enter

Cut_Feed	600
Plunge_Feed	60
Cut_Units	IPM
Retract_Units	IPM
Plunge_Units	IPM

and under MACHINE, enter

Spindle_Speed   2000

and under ENTRY/EXIT

Clear_Dist   50.8

    Select

    Simplified

and exit the machining parameters menu.

Before continuing we should say something about the 45 degree cut angle. Anytime you machine parallel to a steep side, a small step_over means the tool would have to advance downward a large distance before making contact with the surface. That means you'd have to specify a very small step_over to machine a smooth surface. That results in a very slow machining process. So, to permit machining parallel to the a surface without having to reorient our part, we specify a cut angle.

Selecting the Milling Surface

We're ready to continue our sequence. Select

Done

    Select Wind

        Find Now

and highlight the name of the window we previously created (MW1). Pro/E picks all the surfaces that can be machined in the window with the current operation. Select

    OK

OK

Play Path

Screen Play

The results look okay at first glance, but there is at least one readily noticeable thing we don't like. The machine process starts in a corner away from part home. While not a real problem, we generally want to start the process close to part home. Let's go back and modify our machine parameters to make that happen. Select

Seq Setup

and check the Parameters checkbox. Select

Done

Set

and modify the cut_angle to 315. (We determined the correct angle by trial and error!!!) You can play the cutter path to observe the results.

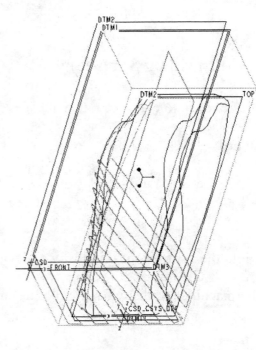

Figure 4-6
315 Degree Cut Angle

NC Check

Now we'll run an NC Check to see what our machining operation looks like. Choose

NC Check

and observe the material removal. It looks like we have everything in pretty good shape, so let's go

back and set our final parameters.

Setting the Final Parameters

We'll set our parameters the way we want them for the output. Since we're already in the

Surface milling sequence, select

Seq Setup

and check the Parameters checkbox. Select

Done

       Set

and modify

Step_Over   1

Exit the Param Tree menu and select

Done

       Done Seq

          NC Sequence

             1:Volume Milling

             Seq Setup

and again check Parameters before selecting

Done

      Set

and modify

Step_Depth   10

Step_Over   8

and we've got it!!!

Generating the Final Output

We're ready to generate the G-code for the entire operation. After exiting the Parameters menu and returning to the machining menu, select

    CL Data

        Operation

and select OP010. Choose

    File

        Done

and enter a file name for the cutter location data or use the default. After the file has been created, select

    Done Output

        NC Check

           Cl File

and select the CL Data file name just created.

    Open

Done

Again, this may take a while.

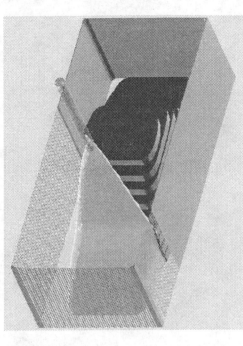

Figure 4-7
The Finish Cut

If you're satisfied with the results, you can again create your CNC file. Select

Output

> **Operation**

>> **OP010**

>>> **File**

and check the MCD File checkbox. Enter the NC file name you want. Select

Done

and the mill description that comes closet to your controller. Pro/E will create your CNC file. The

file may require editing as discussed in Section 2.

NOTES:

Section 5 — Conclusion

In the conclusion, we'll discuss a lot of odds and ends that we just couldn't find a convenient place for discussion in the earlier sections.

Menu Redundancy

We pointed out, and you should have noticed, the redundancy in the menu picks as you worked through Pro/Manufacturing. It is both a curse and a blessing. For the novice user, the redundancy can be very confusing. For instance, you can specify the Manufacturing Geometry while in the Manufacturing Setup or Machining menu. We don't know why PTC decided to do it that way, but it gives the user a lot more freedom — once you've mastered the process.

Tool Parameters

The tool parameters menu, likewise, shows up in different places. This, too, gives the user a lot of versatility. However, if you use the manufacturing module a lot, you'll find it's much more convenient to establish a file of tool parameters for different scenarios. You can have a tool file and a parameter file for roughing, finishing, profiling, drilling, etc. You can also create and save files that have a mixture of English and metric units, as in Section 4 of this manual. In most cases then, this file can be retrieved and used without any editing. Even minor editing is quicker than reentering all the information each time you develop another machining sequence.

Editing Sequences

If you use the Pro/E Manufacturing menu, sooner or later you're going to have to do some major editing of a machining sequence. In the previous sections, we made some changes and did some editing while we were in a specific NC Sequence. You can always go back into a sequence and edit it as we did in the previous sections. But if you want to delete or reorder a specific sequence, the procedure is a little different. Enter the Manufacturing menu and select

Machining

Utilities

and select the operation you'd like to perform. You can Modify, Delete, Redefine, etc. If all else fails, you can delete the sequence and reenter it. If you'd just like to modify the sequence, select

Modify

NC Sequences

and select the sequence that you'd like to edit. Notice that you can modify the tooling, tooling parameters, etc.

Tool Path Files

We'd be negligent if we didn't mention something about the different tool files. As you work with Pro/Manufacturing and work with cutter location paths, Pro/E creates a file with a .tph extension. As you continue to work in Pro/Manufacturing, Pro/E continues to append to this file. It can become very, very large. If you don't delete these files, you're wasting a lot of disk space and may run out. These files can be deleted at any time in the process. As soon as you "play the path" of a cutter, any needed part of the file will be recreated.

When you "save" a cutter location file, Pro/E creates a permanent file with a .ncl extension. This is an ASCII file that can be "viewed" with any text editor. It's also written according to an ANSI standard, so many third party vendor's post processors can use the file to generate G-code. If you're of such a mind, you can use this file and create your own post processor program.

When you use one of Pro/E's built in post processors, Pro/E creates a file with a .tap extension. This is the file that contains the G-codes necessary to program your machine. As we stated earlier in this manual, this is an ASCII file and can be edited for any discrepancies with your specific machine.

Visual Checks

As you play the cutter path and perform NC checks, you can get a good illustration of how well Pro/E is performing the task you've defined. Sometimes, however, you might wish to determine if the rough machining allowance you specified is being left. Or you might want to make sure the finish cut is removing all the material you specified. You can use the tool cutter paths and some of Pro/Manufacturing's features to make a closer observation.

As an example of how to use these features, retrieve the part we machined in Section 1 of this manual. Play the cutter location path. Zoom in on one of the edges or corners of the letter. Reverse and then replay the path. You'll be able to see the tool paths. From the Play Path box, select

Position Cutting Tool

and click on the tool path (red line) in a location you'd like to investigate. Pro/E will display the tool at that location. You can use the Play buttons (Slide the Display Speed all the way to the left) to observe the cutter motion. You can also rotate the view to make sure that Pro/E is removing or leaving the required finish stock.

We can use the same technique to investigate the drilling operation in Section 2. Did you wonder how deep the drill penetrated into our workpiece? The NC check makes it appear as if the drill went all the way through. But how far? Retrieve the manufacturing file for that section and use the Machining menu to recall the NC Sequence for the holemaking sequence. Play the Path. Now establish a front view of the part (so the top surface appears as an edge) by either using Pro/E viewing commands or rotating the view.   Select

Position Cutting Tool

and click on the tool path near the bottom of one of the drilled holes. Pro/E will display the drill at that position. Use the Play button to establish how deeply the drill penetrates through the workpiece. Looks like we need to make sure the machinist sets the workpiece on some sort of backing material.

Conclusion to the Conclusion

That concludes our tutorial. We've tried to present the material so that you can successfully produce G-code to perform some basic machining operations. We hope that you've found the manual to be understandable and useful.

Appendix A — The Basic Feature

The first three sections of this text all use the same basic feature — a rectangular solid. We'll sketch a rectangular section that's 6 inches wide by .375 inches high and protrude it to a blind depth of 4 inches. Although it's simple to create we include the instructions here so that you'll have the correct orientation. Since we assume you have a working knowledge of Pro/E modeling, we won't include much explanation; just keystrokes. We'll use Pro/E's default part template.

Figure A-1
Our Basic Feature

To create the basic feature, select

File

    New

        ⊙**Part**

and enter a name. We used block for the basic feature in Section 1 and plate for the basic feature in Section 2. We didn't specify a name in Section 3. Select

OK

    Click on the FRONT datum plane as the sketching plane. Select the extrude tool icon

     (Extrude Tool)

    and from the dashboard, select

(Create a Section)

Accept the default sketch orientation (Reference: Right and Orientation: Right) and select

Sketch from the Section Dialogue box.

Accept the Right and Top references for dimensioning and select

 (Create 2 point centerlines.)

Close

and create a horizontal and vertical centerline along the Top and Right datum planes. Select

□ (Create rectangle.)

(create rectangle) and use the mouse to create a rectangular box centered about the default

coordinate system. Modify the rectangle to be 6 inches wide and 0.375 inches high.   Select

✓

to continue the sketch. Select

(Extrude on both sides.)

and enter a depth of 4.0 inches.

✓

You've created the basic feature with the correct default view for the first three sections.

Appendix B, C and D contain the remaining steps for Sections 1, 2 and 3, respectively.

Appendix B — Protruded Letter

We'll produce the part required for Section 1 by continuing with the basic feature we created in Appendix A. Note that we have "turned off" the default datum plane and coordinate system display for the section.

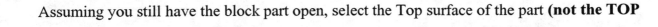

Figure B-1
Part for Section 1

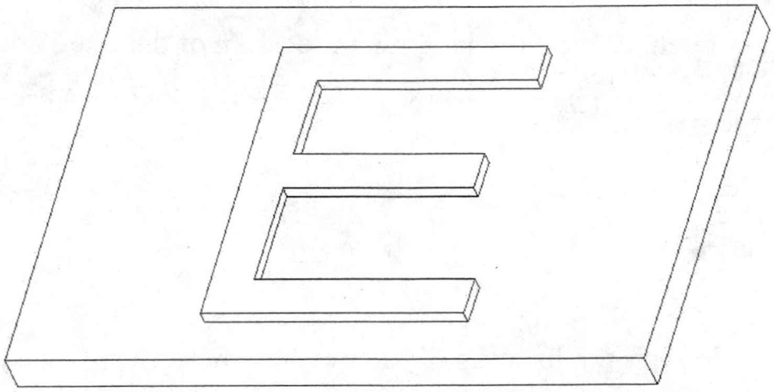

Assuming you still have the block part open, select the Top surface of the part **(not the TOP datum plane**; Refer to the Basic Part Orientation on page iv-1.) as the sketching plane. Select

 (Extrude Tool)

 (Create a section.)

Leave the default references and select

Sketch

Close

 (Create Text as part of a section.)

and sketch a vertical line on the rectangular section. The line should **start** about one half inch from the **bottom edge** of the rectangular section and extend to about one half inch from the top edge.

Create the line about a third of the distance from the left edge to the right edge. When you finish

sketching the line, Pro/E will display a dialogue box for entering text. For this example, you

probably want to use the text we've suggested. A different character might require a smaller cutting

tool then suggested in this manual. Enter a capital E. You can adjust the location and height of the

text by modifying the length and location of the line. You can modify the width by adjusting the

aspect ratio. Select

and modify the location if you need to. Select

and enter a depth of 0.25 inches. Select

and you've completed the part required for the first section of this manual.

Appendix C — Part for Section 2

We'll produce the part required for Section 2 by continuing with the basic feature we created in Appendix A.

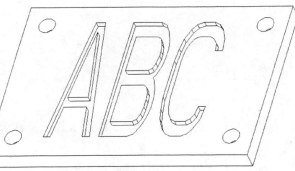

Figure C-1
Part for Section 2

We'll start by creating the four holes. We'll create one hole and pattern it to produce the other three holes. Assuming you've already created the block part and have it open, select the Top surface of the part (not the TOP datum plane; Refer to the Basic Part Orientation on page iv-1.) Select the hole tool icon

(Hole Tool)

Drag the horizontal hole handle to the top-left edge of the part. Drag the vertical hole handle to the top-front edge of the part.

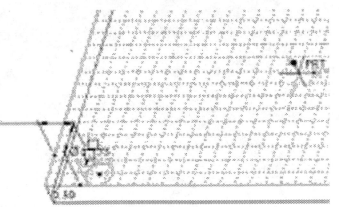

Figure C-2
Locating the Hole in the Plate

Select the "Drill to intersect with all surfaces" icon from the dashboard.

 (Extrude to intersect with all surface.)

Select

to complete the first hole. With the hole still highlighted, select

 (Pattern Tool)

and pick on the 0.5 dimension that is parallel to the left edge. Click the right mouse button

somewhere on the screen and select **Direction 2 Dimensions**. Click on the Dimension Tab (Top

left hand corner of the dashboard just to the right of the Pattern Tool Icon that is being displayed.

Enter a Direction 1 increment of 3 and a Direction 2 increment of 5.0. Select

And we've created our 4 holes.

We now need to produce the cut out of the lettering. Referring to the default orientation at

the beginning of this manual (Figure A), select the top surface of the basic feature (top surface of

the block). Select

(Extrude Tool)

(Create a section.)

Accept the Sketch Plane and Reference. Select **Sketch** from the Section Placement Dialogue box.

Close the references box and select

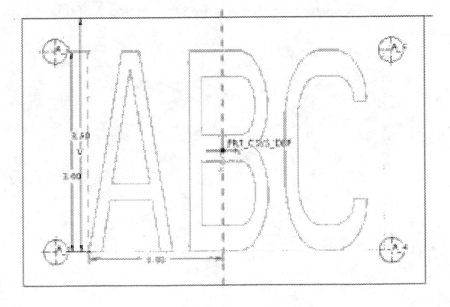

(Create text as part of a section.)

and enter a vertical line **starting at the bottom** of the rectangular section extending vertically

toward the top. The line should start about ½ inch from the top-front edge of the part (Refer to page

iv-1. It should be on the bottom of your screen.) and extend to within ½ inch of the top-back edge.

It should be located about ½ inch from the holes on the left side of the part. Enter the text to be

machined. For this example, you probably want to use the text we've suggested. Enter ABC. The

letters will probably extend outside the boundaries of the section. Use the aspect ratio to adjust the

letters to fit between the holes with ample clearance on each side. Accept the text dialogue box.

(You can display the box again by clicking on the "Modify" icon and selecting the text.) You can

modify the dimensions to change the location of the text and change the aspect ratio again if

necessary.

Figure C-3
Letter Location

Select

Done

and enter a depth of .125 inches for the depth on the dashboard. Select

(Remove Material)

(Change depth direction)

to remove material from the part. Select

and you've completed the part required for Section 2 of this manual.

Appendix D — Part for Section 3

We'll produce the part required for Section 3 by again continuing with the basic feature we

created in Appendix A.

Figure D-1
Part for Section 3

Begin by selecting the top surface of the part as the sketching plane. Select

 (Sketched Datum Curve Tool)

and pick on the front surface of the basic feature for the reference orientation (**Bottom**). Select

Sketch

Again referring to the default orientation, select the top-left edge and the top-front edge for

references for dimensioning. Close the references dialogue box and from the right toolbar, select

Close

(Create a spline curve.)

Use the mouse to produce a "click by click" spline for the character to be machined. Don't be too

particular. After you've finished entering the spline, you can modify the points and drag them to

new locations. To terminate spline creation, hit the middle mouse button. Pro/E will dimension the

beginning and end point of the spline curve you've just created.

An interesting variation of what we've used here is to write your name in script. Be aware that any break in the spline will require another feature creation. Also, in Section 3 we suggest using a 1/8 inch milling cutter which requires that you not make any section to be cut out too narrow or the lines will overlap and be indistinguishable. When you've finished sketching you can modify the spline using

       (Modify the values)

and dragging points to new locations. You can also modify the dimensions of the end points. But be careful redimensioning the endpoints. If all your changes are not proportional, you can alter the shape of the text. Pro/E will do mathematical calculations so you can use a scale factor on each dimension if you wish. Select

and you've completed the first letter.

The same procedure can be used to create the second letter.

Appendix E — Part Model for Wooden Car

In this appendix we'll show you how to create the design part for Section 4.

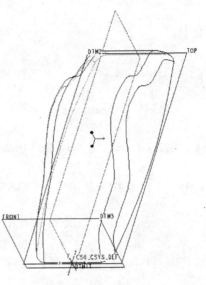

Figure E-1
Part Model for Section 4

We won't start with the basic feature as in the other sections. Our basic feature will be the default

datum planes and coordinate system in Pro/E's default part template. Since we're creating the car

in metric units, we began by setting millimeters as the default unit of measure. Select

   File

      New

         ⊙**Part**

and name the part car.prt.

   OK

To define millimeters as the default system of units, select

   Edit

      Set Up

Units

and select "millimeter Newton Second (mmNs)".

Set

OK

Close

Done

The basic outline of our car will be created by sketching a spline curve on the default datum planes. We'll protrude this curve to produce one half of the basic shape of the car. We'll make a vertical cut using an arc to "round" the car so the front and back are not as wide as the central section. The sides will have a slight taper inward from bottom to top, and the outside edges will be rounded. A mirror operation will complete the model. We begin by creating a spline protrusion on the default datum planes. Select the Right Datum plane as the sketching plane and select

 (Extrude Tool)

(Create a Section.)

and in the Section Dialogue box, make sure you have the following:

Plane: **RIGHT**

Reference: **TOP**

Orientation: **TOP** (This one will probably have to be changed.)

Sketch

Close

(Create a spline curve.)

and sketch a cross section of the car. Place the bottom front of the car at the origin of the default coordinate system and sketch the car cross section to the right. After you finish sketching the cross section, you can use modify to move around the points to adjust the cross section to your satisfaction. Click the middle mouse button to terminate the spline creation.

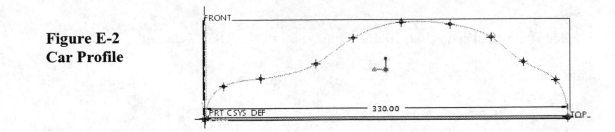

Figure E-2
Car Profile

Use

(Create 2 point lines.)

to place a horizontal line along the bottom of the car from front to back.. Dimension the car for an overall length of 330 mm. Select

(Modify the values)

and modify the outline of the car to your satisfaction by dragging around the points. Make the car height approximately 25% of the length.

and either drag the depth handle or enter a blind depth of 120 mm.

We'll now make the car width thicker at the middle than the front or back. After selecting the TOP datum plane, select.

(Extrude Tool)

(Create a section)

Sketch

> **Close**

> (Create 2 point centerlines)

and place a vertical centerline 50 mm to the right of the RIGHT datum

plane. Select

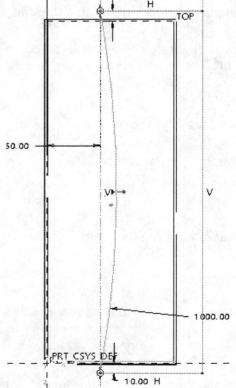

> (Create an arc)

and place an arc that begins and ends on the centerline you just created

and extends 10 mm outside the profile of the car. Dimension the end points of the arc (the 10 mm)

and dimension the radius to be 1000 mm.

Figure E-3
Arc for Cut

Select

> (Create 2 point lines.)

and create a rectangular section that encloses the remainder of the car as shown in Figure E-1.

e-4

Select

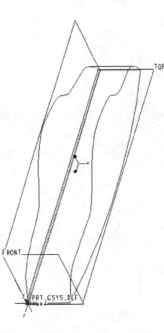

(Extrude to intersect with all surface.)

(Remove Material)

make sure the arrow indicates the proper material is being removed and select

Figure E-4
Car After Circular Cut

We'll put a draft on the side of the car so the top width will be narrower than the bottom.

Select the curved-side surface of the car we just created and select

(Draft Tool)

Select the References tab and select the "No Items" choice under Draft hinges. (Note that a surface

has been selected under Draft surfaces.) Select the TOP datum plane as the draft hinge. Use the

draft handle to place a 10 degree draft on the side of the car – wider at the bottom then top. Select

We'll now place round on the curved edge of the car before mirror it to produce the complete car. Select the curve representing the car profile (the one we just place the draft on) and select

 (Round Tool)

and enter a radius of 15 mm. Select

![checkmark icon]

We've completed one half of the car. If we mirror this half, we'll have the complete car. From the model tree, highlight the part node (car.prt from the very top of the model tree). Select

Edit

    Mirror

and select the Right datum plane as the plane to mirror about. Select

![checkmark icon]

THE CAR MODEL IS COMPLETED.

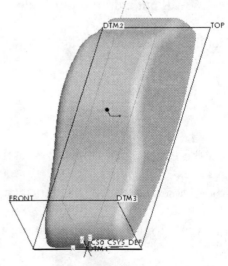

**Figure E-5
Completed Car**